Paradox of the Learning Game

The Promise and Plight of Video Games and Learning

Marcus T. Wright

Edited by Alison Perch
Cover and Illustrations by Bola Onayemi

design
meets mind

Printed in the United States of America
First Printing, February 2017
ISBN #: 978-0-9985570-0-7
Design Meets Mind, LLC: Philadelphia, PA

CONTENTS

To Channy. Your support means the world to me.
And to August. You make us both better.

Paradox of the Learning Game

A QUICK WORD...

Modern technology is a gift and a curse. I do not think it is possible for us to really understand its power and potential in this current generation (unfortunately). It will be quite some time before we can fully grasp how to best leverage computer and digital technology for the good of society. On the road to that understanding, there will be frustration and discouragement about what we *cannot* accomplish with technology, or how technology does not always come through for us in the ways we would like it to. I believe this frustration and discouragement is rooted in the idea that "technology is the answer." Thus, if we do not get the "answer" from a technological innovation, or if we do not get that answer quickly, we swiftly move on to the next "great" idea.

I think this is a misguided train of thought. Rather, we should look at technology as a *gateway to possibilities*. Every technological advance and innovation opens the door to new ways of doing, new ways of making, and sometimes new ways of *being*.

That is the idea that underlies this book, which I like to consider as a brisk foray into the technology we fondly call the video game. There are many people who are exploring how we can best extend the use of video game technology beyond entertainment. Particularly, there has been an influx of discussion on whether video games can effectively facilitate knowledge that can be transferred to real-life contexts. In this book, I will discuss games made specifically for this purpose — *learning games*. These games have garnered steady attention in recent years, yet there is still so much we do not know about their benefit to human learning.

Learning games are a prime example of how we cannot conclude that technology is the answer—at least not yet. That is okay. Instead, we should focus on the various possibilities that are created with every learning game. In order to do this, we must critically analyze learning games as a whole; not just their design and methods of assessment, but also their impact and place in society. This book is my attempt to do so by looking at video games as a *medium*, and subsequently analyzing the challenges that emerge when using this particular medium to facilitate the complex phenomenon we call learning.

Thank you so much for picking up this book. Whether you are for games or against them, and whether you research games or forego much needed sleep to play them all night, I believe there is something in this book for you. Let's keep moving toward a future where we better understand what technology of all varieties can truly do for us.

-*Marcus*

Level 1

Learning Technology of the Future

Say hello to Hero, the Knight.
He's the...uh, hero...of the story.

"And in the same way that we invested in the science and research that led to the breakthroughs like the Internet, I'm calling for investments in educational technology that will help create digital tutors that are as effective as personal tutors, and educational software that's as compelling as the best video game."

-President Barack Obama during a speech at TechBoston Academy in Boston, Massachusetts (The White House, 2011).

***

In 2012 I became an advocate for the power of video games for learning purposes. It was a nice epiphany, which I will detail a bit later in Level 1. Ironically, I had this revelation at a time in my life when I *seldom played video games*. Sel-dom.

I know that books on the benefits and positive potential of video games usually start with the author saying something along the lines of, "I'm playing a video game *as I write this book!*" Well, not this time.

I certainly grew up playing video games all the time. Binging on video games was a norm of social life for me, my peers, and relatives. Video games took us to worlds unknown and challenged us to overcome complex obstacles. They allowed us to help fictional characters overcome ridiculous challenges and complete fantastic journeys. To me, video games were portals to the peaks of imagination.

Somewhere along the way, life started doing what life does, and I had to drastically lessen the amount of time I spent playing video games. The decline was most noticeable to me after I completed my undergraduate studies at Rutgers University in 2006. I guess then, that I should blame the whole "entering the real world" thing.

Even though I played video games less than ever before, I remained a fan of the industry. I followed the latest news and updates whenever possible. Still, I had transitioned into a casual gamer, only playing games occasionally whenever I had the audacity to tell my responsibilities to "stand back!"

I still consider myself to be a casual gamer today. Consequently, there is no way I could present this book to you as though I log hundreds of hours saving worlds and customizing characters. However, even though the real world has valiantly dampened my aspirations of overplaying video games, my fascination toward the *video game medium* is greater than ever. By this, I am referring to the use of video and digital technology to provide a carefully designed, interactive, and playful experience to an end user. These experiences normally come in the form of what we know as the *video game*. These games are presented to us through various platforms such as a disc for a home console, a digital download from a website, software for a computer, or a mobile application (see the table on the next page).

Much like print, film, and other media, video games have emerged as a way for people (those who collaborate to create each game) to capture the imagination of other people around the world. However, I believe that the video game medium *needs* people on a more complex scale than other media. The byproduct of the video game medium—the actual video games—are unfinished ventures, awaiting the input and participation of one or more players. Without these players, video games can only be beautiful visual and audio exhibitions through their demonstration modes. If this were the case, the creators of the game could have just made a film!

VIDEO GAMES – THE GAME, TECHNOLOGY, AND MEDIUM

VIDEO GAME	VIDEO GAME TECHNOLOGY	VIDEO GAME MEDIUM
A carefully designed, audiovisual product meant to facilitate an interactive and playful experience for one or more players	How the video game is accessed, (e.g., disc, digital download) which in turn will dictate the complexity of graphics, audio, etc.	The use of video games to create an interactive, playful experience for an end user (or users). This experience will be influenced by player choices and game strategies
Example: *The Sims, Call of Duty*	Example: Sony PlayStation Discs + console, or in the case of computer games, software or digital download + computer	Example: A team of designers putting together a game that puts players in the shoes of characters surviving a zombie outbreak; the outbreak was caused by a virus (the *Resident Evil* series)

The video game medium enables game creators (designers, writers, composers, and anyone else who works on the game) to craft a unique and interdependent relationship between video game and player. Game designers create level after level and challenge after challenge with the input of potential players in mind.

> **Thus, the only way a video game can become a complete product is if someone actually plays the darn thing. This is a mantra that will always apply to video games.**

A book can sit on a shelf and still consists of its pages and story, and a film can be played in front of an empty movie theater and still run from beginning to end. A video game that is never played, however, is incomplete because it was designed to be progressed and completed by player input and participation.

This is not to diminish the need for people when it comes to other media, as books and such are created with the notion that the message within them will be received and interpreted by end users. As a form of media, video games share this characteristic. Also like other media, video games generally have a beginning, middle, and end to the storylines within them. Yet with games, most aspects of what the journey will look like through that story hinges as much on the receiver of the message (the player) as the game itself.

One modern outcome of this interdependent relationship between video game and player is the significant sense of catharsis that people experience by playing and following video games. This goes beyond the reality that we invite videos games into our homes, social circles, and everyday lives to play them and watch others play. The capabilities of the video game medium, along with the various elements of the video games themselves (such as the artwork, characters, background music, level design, storylines, etc.) has created a social phenomenon that routinely penetrates the boundaries of mainstream culture.

This is why people play video games against others with large groups of people watching[1], listen to video game soundtracks online, or create custom game cover art to share on social media. This is why people cosplay as their favorite video game characters (comic convention or not) and extract cut-scenes from video games to form mini-movies online. There is so much that we can do with a video game, because the medium inspires these opportunities.

Because of this, the impact of video games has defied social parameters. These games shape the everyday lives of many people in our society. This has been magnified in recent years thanks to the rise of computer and digital technology, which has widened the scope of what can be accomplished through the video game medium. Today we see video games with increasingly realistic graphics and game worlds that are as engaging and responsive as ever. Even games with less complex graphics (or even just words, as in the case of interactive fiction[2]) are reeling in legions of players with imaginative gameplay and storylines.

---

[1] This refers to eSports, an increasingly popular industry, where gamers from all over the world gather and compete against others (individually or as part of a team) in front of large audiences. According to market data from SuperData Research, Inc. (2016), eSports revenue is expected to reach $1.1 billion by 2018!

[2] Interactive fiction games focus more on story as opposed to the visual, so these games are often characterized by text only. Players can type in commands or select from various options to progress through the game.

It's jarring to consider how far video games have progressed in such a short period of time. One of the most well-known and original commercial video game titles, *Pong*, was released in 1972[3]. In less than five decades, we have come from the days of a black-and-white game with two lines hitting a ball across a screen, to the immersive creations of today that dominate the time—and wallets—of people of all ages. In fact, in 2015 video games, computer games, and games delivered through other platforms comprised $16.5 billion in sales in the United States—a 7.1% increase from the $15.4 billion in sales in 2014 (Entertainment Software Association, 2015; Entertainment Software Association, 2016).

All of this is due to the capabilities of the video game medium, not any particular video game itself. It is important to recognize this distinction, because it forms the basis for any conversation about what *we can do* with video games. The power of the video game medium goes beyond the constraints of any single piece of video game technology. For example, *Pong*, as seminal of a game as it is, is one game. It is the video game medium, however, that has captured the imaginative spirit of players to the point that we see spectacles such as *Pong* being played on a skyscraper in Philadelphia![4] Truly, the medium lends itself to the creative whimsy of people across the world. This is why video games (once again, the byproduct of the medium), whether computer, console,

---

[3] Release date as listed on Atari's official website (Atari, Inc., n.d.).

[4] In April 2013, gamers gathered at the Philadelphia Art Museum to play Pong using the side of the city's 430 foot skyscraper, the Cira Centre, as the screen! This endeavor, led by Drexel University Professor Frank Lee, was billed as "The Grandest Game of Pong on the Planet" (Wink, 2013).

mobile, or digital, are an extremely important technology in the world today.

For this reason, I believe it is important to consider all of the affordances of the medium for the benefit of people and society. We already know that video games bring entertainment into our lives, along with opportunities for social bonding and even the appeal of spectacle. This may only be the tip of the iceberg. The video game medium is unlike any other: cloaked with the ability to create a unique, interdependent relationship between itself and the end users (players). With this in mind, game designers, scholars, and educators are eagerly pursuing the question:

> ## What are additional ways that we can use the video game medium to make lives better?

Now on to that nice epiphany I had in 2012.

THE LEARNING EPIPHANY

As someone who has worked extensively in education in and out of the classroom, I find the complexities of learning—how does it happen and why—quite interesting. We learn all the time, everywhere we go. Most of this learning happens so fast, and is so deeply embedded in our daily routines, that we do not even realize that we are learning.

New technologies have thrown the ultimate variable into this whirlwind. We can actively use technology to help us learn (such as an on-the-spot internet search on how to do something), or technology can be used to mediate knowledge to us (such as a video display at a museum exhibit). There are just about an infinite number of ways that the *learning and technology* dynamic can manifest. This has motivated me to explore the relationship between these two phenomena for the past several years.

Under this motivation, I entered the Learning Sciences and Technologies Master's program at the University of Pennsylvania's Graduate School of Education (Penn GSE) in 2011. I intended to explore how technology could be used to create engaging learning environments within educational institutions. I wanted to answer the question: how could we as a society best use technology to facilitate learning for individuals and groups? In turn, how could we best use the so-called "learning technologies" of the world today?

In retrospect, these were—and always will be—dense questions to pursue! Nonetheless, I rigorously approached these questions during my years of study. The lightning pace of technological innovation during those years did not make things any easier. There was always a flood of learning technologies vying for the public's attention. Unfortunately, the arrival of these new technologies rarely seemed to be paired with thorough critical analysis. In fact, it seemed that hyperbole was routine, as each new learning technology had to compete with the hordes of other (sometimes similar) learning technologies. Also, the strong predilections of most stakeholders in the learning technology conversation (including myself at times) seemed to leave little room for objective conversations that furthered the understanding of how to best use these technologies for learning.

One prime example was the emergence of Massive Open Online Courses (MOOCs). This technology enables college instructors (or anyone, really) to teach free online courses with open enrollment to large numbers of students. This development in technology meant that people from all over the world could instantly take courses from acclaimed higher education instructors. The rise of MOOCs was fascinating, and I certainly thought they presented a great opportunity to push the envelope of learning. At the same time, I knew that the biggest thing educators and researchers needed to do was ask questions about the emerging technology. In a 2013 article for *The Huffington Post*, I wrote:

> *"While all of the debate around MOOCs is certainly exciting, we as educators, researchers and educational leaders must be careful not to get too caught up in the art of debating. Instead, we must remain focused on asking the questions that will help us shape whatever we have here with MOOCs into something that will help the most learners possible"* (Wright, 2013, para. 2).

I was excited to study communication challenges within MOOCs for my Master's thesis at Penn GSE. I considered this my attempt to add to the analysis of the technology. The hype for MOOCs, however, was too much too soon, and opened the door for anti-MOOC sentiment to rise just as quickly. A 2012 *New York Times* article, "The Year of the MOOC" was often cited in other articles and thought leadership on MOOCs (Pappano, 2012). Although the article never actually declares that 2012 should be considered the year of the MOOC in a definitive, victorious way (and actually provides a thorough analysis of some pitfalls of the technology), the headline was all that was needed for the hype on MOOCs to skyrocket. Naturally, the pessimism to counter this hype grew as well.

Thus, as soon as a study from the University of Pennsylvania found that few MOOC registrants actually completed their courses, and engagement in the courses decreased after the first two weeks, the air quickly let out of the MOOC balloon (Perna et al., 2013). While MOOCs are still around today, it seems that many advocates have jumped off the "hype train," and I'm not sure if the technology will be able to make the same impact that it would have if we had only stopped and *asked more questions*.

I digress.

Amidst the sea of such rhetoric for MOOCs and other learning technologies, I found it difficult to determine which of these technologies were genuinely effective and what methods were best for integrating them into learning environments. I went through mounds of thought leadership, research and analysis on this subject, but nothing grabbed my attention as a true breakthrough for learning.

That changed in 2012, when I took a course at Penn GSE called *Video Games and Virtual Worlds as Sites for Learning* (created by my program advisor Dr. Yasmin Kafai and taught that semester by Dr. Deborah Fields of Utah State University). When I signed up for this course, I thought:

> **"Hmm...learning in video games?**
> **Seems cool and it fits my schedule.**
> **Why not?"**

I showed up to the first class not really expecting much. So, of course, this was the graduate school class that completely blew my mind! The course helped me realize that video games are more than just tools for fun and entertainment; they also foster a variety of learning experiences within them (I will detail this in Level 2).

The very thing that I always considered to be the embodiment of anti-learning was in fact a beacon of learning.

Go figure.

When I first entered that program in Penn GSE, I would have laughed at anyone who said, "You will come out of this program thinking that the video game is one of the most promising learning technologies in the world." Yet that is exactly what happened. Even though I finished my studies focusing on MOOCs, I just could not shake the feeling that there was something great about the concept of *video games and learning*.

That brings us to today. I still have that feeling, more than ever. I believe that the video game, in its various forms, has the potential to be the "learning technology of the future." The possibilities that can be created with each game are far too great to ignore. The affordances of the video game medium enable this potential, which has been magnified by today's technological advances. I feel that the video game medium has the capacity to handle the complexities of learning unlike any other medium today. The key is to (all together now…) ask the right questions along the way.

There's my hyperbole. However, I want to engage in some critical analysis as well! In particular, I want to focus on video games that have been built with transferable learning[5] in mind — or *learning games*. These games facilitate learning experiences with the expectation that players will draw from (or at least *could* draw from) these experiences in real-world situations. For this book, I

---

[5] For this work, by "transferable learning" I am referring to knowledge that is facilitated in video games that is meant to be used by the player in real-life situations if needed. For example, a math learning game may have the player learn algebraic equations in order to advance in the game. The ultimate goal of the game is for the player to learn those equations well enough that he or she will be able to solve such algebraic equations should they appear in real-life.

am referring to games that develop academic skills (such as math and science) and soft skills as well (such as emotional intelligence, collaboration, and social awareness).

Although many types of games use the learning game label (or its infamous alias, *educational game*), too many of them only marginally take advantage of the affordances of the video game medium. They may consist of tasks to complete, flashy graphics, background music and boisterous sound effects, but they fail to ask the player to become immersed in the type of game worlds, narratives, and consequential decision-making that can be found in entertainment games. The term "chocolate-covered broccoli" has often been used to describe these games. This not-so-flattering term is defined in the 2013 Open University Innovation Report as an approach where the game "provides a veneer of fun covering a mundane educational task" in which the "game may offer a stimulus or reward, but the underlying exercise does not change" (Sharples et al., 2013).

I maintain that this type of game was sufficient in the past when the video game medium was younger, and the scholarly efforts to figure out how to best combine engaging gameplay with critical learning was minimal. Yet, given what we know today about technology, game design, and learning, creators of modern learning games should be able to avoid the chocolate-covered broccoli approach and provide games that significantly take advantage of the video game medium. However, creating such a game is not an easy feat, which could make developing a game or application with the "veneer of fun" covering over educational tasks a more attractive option.

Thus the learning games that excite me, the ones that I advocate for in this book, aim to emulate gaming experiences that rival what can be found in entertainment games on video game consoles, computers, and mobile devices. They ask players to engage in game worlds with critical decision-making and problem solving strategies (more on this in Level 2). They also dare to ask players to achieve learning goals in creative ways that are intricately tied to the gameplay and narrative. Flashy graphics, catchy background music, and boisterous sound effects can be elements in these games, but are not necessary. These are the learning games I believe have the most potential for transferable learning, as they take significant advantage of the interdependent relationship that is created between video game and player. Operatively, because these games include more advanced usage of the affordances of the video game medium (regardless of platform and game length), I will refer to them as *advanced learning games.*

Advanced learning games include—but are not limited to—*serious games.* Serious games have been defined in numerous ways, from "games for learning" (Serious Play Conference, 2015), to games that improve "knowledge, skills, or attitudes in the 'real' world" (Graafland et al., 2014), to "games with more than just entertainment as a goal; that embody the principles of deeper learning" (DiCerbo, 2015, para. 3). Although the definitions vary across the board, there is a consistent theme that serious games are distinct from games made strictly for entertainment purposes.

The label of "serious" is a bit tricky, however. I believe that defining games with learning intent as serious games inadvertently positions games without explicit learning intent as *not* serious. By default, that weakens the video games and learning conversation because we position the very foundation of the concept—video games—as not serious unless we add learning to it. Perhaps I am looking too deep into everything, but because the video game medium is such a fantastic entity, I think that *all* video games are serious—whether they include transferable learning goals or not. Therefore, I will use the term advanced learning game throughout this work, and under this term I include any title that is considered a serious game.

THE PROMISE AND PLIGHT

We are used to the idea that learning games are best fit for the educational classroom—and rightfully so. Educational institutions harbor learning experiences that are meant to be drawn from in real-world situations. Any technology that can support this sort of learning becomes a valuable resource, especially something as popular with kids, teens, and young adults as video games.

Yet we should not underestimate the potential value of these kinds of video games for lifelong and anytime/anywhere learning outside of the classroom as well. These forms of learning are particularly important in our competitive 21st century. People of all ages are doing whatever they can to learn skills and earn credentials outside of structured educational environments for their academic and professional portfolios. Advanced learning games could be great tools for these scores of people—but only if these people have access to, are aware of, and actually get to play these games (remember, video games are incomplete without players). In my ideal future, designers and publishers of advanced learning games would have no problem finding expansive audiences, and audiences would have no problem finding, learning from, and enjoying advanced learning games.

There are many challenges to reaching this kind of future, which I will detail further in Level 3. For example, there has been much discussion on the problem of design and assessment; how do we *actually know* that players are learning from these games? Then there is the problem of monetization; how can we sustain advanced learning game production when video game production can be quite expensive? While these are extremely important parts of the video games and learning conversation, there are other factors that need to be considered in order to create a bright future for advanced learning games. These factors, which seem to fly under the radar most of the time, are connected to the place in society that learning games have in general.

There are several obstacles that work against the public awareness and visibility of advanced learning games simply because of the fact that they are *learning* games. It is a paradoxical existence that throws a wrench into the vision of using video game technology for the purpose of transferable learning. If we are ever going to reach a future where advanced learning games are considered vital tools beyond the classroom, we must address the incompatibilities that are produced when the complexities of learning are married with the wonders of video gaming. This is because advanced learning games fall out of the realm of "quick-fast-in a hurry" learning that we can get in many learning games, and almost by default become competitors for the eyes and ears of gamers who *could* be playing other types of games (especially of the entertainment variety).

I choose to look at a few of these conceptual conflicts through my background in sociology, communications, and the learning sciences. I start in Level 2 by detailing the learning that happens within video games, and describing how that dynamic has led to excitement over the possibilities of using the video game medium for transferable learning. In Level 3, I will explore the current state of learning games and detail some examples of advanced learning games. Levels 4 and 5 break down two paradoxes[6] for learning games in general. "The Video Game Paradox" deals with how advanced learning games benefit from and are disadvantaged by the fact that they are video games. "The Learning Paradox" deals with our society's love/hate relationship with the concept of learning. These paradoxes ultimately affect how successful advanced learning games can be.

Levels 6 and 7 explore two tensions[7]; these are not quite paradoxes, but they highlight how learning games are at odds with certain social trends. "The Tension of Failure" discusses our society's impetus to punish failure and how that contradicts one of the strengths of video games — the freedom to fail. "The Tension of Educational Gaming" talks about the impulse to create educational versions of entertainment games and how this shortchanges what advanced learning games can accomplish.

Through this work I hope to contribute to the important conversation on a learning technology that has boundless possibilities. To enable the brightest future for advanced learning games, it is necessary to ask the right questions and thoroughly analyze their place in society. In essence, this book looks at the core of whether advanced learning games can really work in the type of society we live in.

The scope of this work is not all-encompassing. As I mention in "A Quick Word…," this book is a brisk foray into the technology we fondly call the video game. I aim for this to be a conversation starter, but understand my limitations. I bring forward an analysis

---

[6] By "paradox," I am referring to how certain strengths of learning games can simultaneously be considered weaknesses, thus canceling out the benefits that the strengths would have provided.
[7] By "tensions," I am referring to the conceptual conflicts that can emerge due to the learning game's place in our particular society.

of some potential challenges for the future of advanced learning games, but can only offer suggestions of avenues to take to begin addressing those challenges. It is my hope that this work can serve as a call to look deeper at the incompatibilities between our society and learning games in general, and how this affects the future of learning games that take significant advantage of the video game medium. From there, others can approach this topic with the perspectives and resources where I fall short.

In the meanwhile, this is my opus on video games and learning. If we can at least bring paradoxes and tensions such as the ones in this book to the forefront, we can begin to build a stronger foundation so that advanced learning games of all types—not just the occasional hit—can thrive.

Level 2

The Concept of
Video Games and Learning

It would probably be a good idea to discuss what the big deal is with the "video games and learning" concept in the first place, right? Perhaps you are like how I once was: under the belief that "video games" and "learning" should never be said in the same sentence, unless that sentence was, "I don't feel like learning; I'd rather be playing video games." But lo and behold, there is a connection, and it's one that is pretty much *unavoidable*.

One scholar who has rigorously analyzed this connection is Literacy Studies Professor James Paul Gee (Arizona State University). Gee is widely regarded as one of the most renowned advocates for the power of video games as resources for learning experiences. His book, *What Video Games Have to Teach Us About Learning and Literacy* (2003) is often referenced in any work on video games and learning. In a 2011 interview with Digital Media and Learning Research Hub, Gee laid out the fundamental core of a video game:

> *"What is a video game? It's just a set of problems; it could be anything. Civilization is problems in history. Halo is problems in the fantasy world of fight. Chibi Robot is problems of how a four-inch robot can clean a house. Doesn't matter what the problems are. All a video game is, is a set of problems that you must solve in order to win"* *(Digital Media and Learning Research Hub, 2011).*

This perspective — that video games consist of sets of problems that need to be solved — positions these games as significant sources for continuous learning.

Gee's stance on video games is parallel to what we face in real life: we encounter problems on a daily basis, and we try to figure out how to solve them in order to get past them. Sometimes we can solve these problems by recalling helpful information from our past. Other times we need to learn new information, skills, and processes to get through these problems. This is the case in video games; as a player comes across new challenges and obstacles within the game world, he or she will have to either recall how to overcome these obstacles (from something previously learned, most likely in the same game) or learn a new strategy to overcome them. As outlined by Gee, video games are chock-full of this problem-solving dynamic.

Professor of Informatics Kurt Squire (University of California, Irvine), another renowned voice in the video games and learning conversation, highlights this dynamic in a 2011 interview with the Connected Learning Alliance regarding video games:

> *"[Games] actually build skills; you have to be good at a game. People who play games develop pattern detection skills; they develop problem-solving skills. You don't get very far in a game if you just think about doing one sort of solution path. You got to kind of step back when you reach problems, look at what is the most ideal kind of solution, and then given what your skills are, what is the best way to go forward" (Connected Learning Alliance, 2011).*

Squire aptly notes the adaptability and flexibility a player must have to problem-solve. This ties into the learning process; if a player recalls a certain strategy to get past a problem in a game and it does not work, that player must be open-minded enough to try another strategy and even learn some new things along the way.

GAME HELPS THE PLAYER

I feel that the existence of this exciting problem-solving dynamic in video games would be irrelevant, however, if the problems *were impossible to solve*. There is learning that occurs during the course of problem solving that helps people inch closer toward actually solving that problem. Once the problem is solved, that is usually confirmation that those people learned enough to get past the problem. In regards to video games, if a player doesn't get any closer to solving a problem in the game world despite the number of times he or she plays the game, how do we know the player is actually learning?

Therefore, an important part of the learning process that occurs in video games deals with a player being able to advance through and past problems—obstacles, puzzles, boss characters, and the like. While a player's skills in the game have a critical role in this advancement, the design of the game is crucial as well. Remember, a video game and its players are joined together in an interdependent relationship. This means that *both* video game and player have to bring something to the experience to facilitate problem-solving in the game world. The problems in a game could be in a form that the player has never encountered before, or has had difficulty with in the past, so the player may need some help learning how to overcome these problems.

Thankfully, video games are typically built with these potential impediments in mind. The 2015 version of the Open University Innovation Report notes:

> *"Game designers embed opportunities for incidental learning within computer games by setting challenges and offering rewards, as well as by providing landscapes to be navigated, rules to be inferred, and the motives and actions of game characters to be interpreted"* (Sharples et al., 2015).

In other words, video games are designed to help players solve the problems within them by helping players learn whatever they need to know to successfully explore the game world as a whole. This alone makes video games some of the most effective learning environments that can be produced!

In my 2014 interview on game-based learning for *The Huffington Post* with educational game designer Dr. Lucas Blair (CEO of Little Bird Games), he states a similar notion:

> *"Games are simply play with rules and they are a great way to learn. The rules in games let us scale difficulty to keep players challenged and facilitate goal-setting to keep them engaged. Games are also excellent for giving learners feedback right when they need it to maximize impact"* (Wright, 2014b, para. 8).

Taking all of these points into consideration, it's safe to say that video games contain an intricate interplay of many factors that help players get from point A to point B. The comprehensive design of a video game works in tandem with the player's skills, leading to (hopefully) problem-solving progress — a.k.a. learning — for the player.

***

LET'S LOOK AT THE PROBLEM-SOLVING AND LEARNING
RELATIONSHIP IN A VERY STRAIGHTFORWARD EXAMPLE:
A PLATFORM-ADVENTURE MODEL THAT HAS BEEN USED
COUNTLESS TIMES IN VIDEO GAMES.

→ Start

In this example, *Adventure Game* is a side-scrolling game where you control a swashbuckling knight named Hero. He is trying to save the world from evil knights. What exactly makes these knights evil? I do not know.

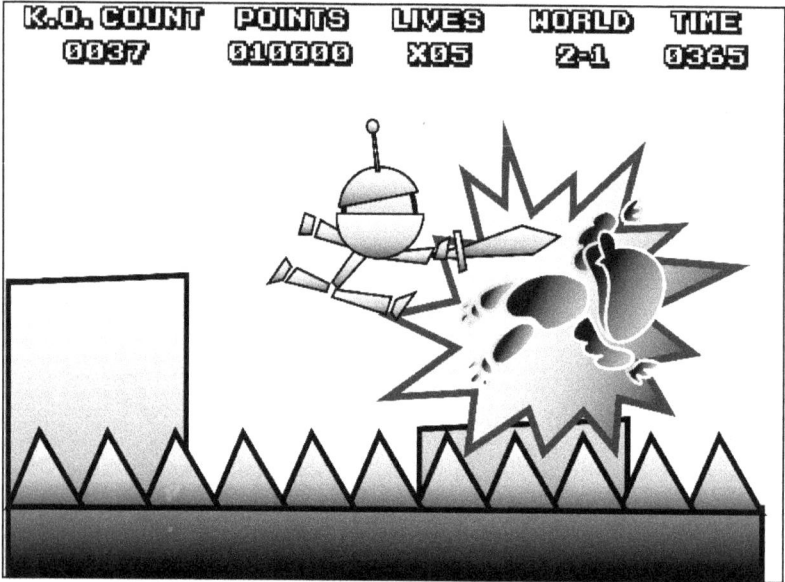

K.O. COUNT	POINTS	LIVES	WORLD	TIME
0037	010000	X05	2-1	0365

- Hero can jump on platforms and on top of enemies to destroy them. He can also use his trusty sword to eliminate the evil knights!
- Every time he destroys an enemy, he is awarded points. For every 10,000 points Hero earns, he receives an extra life.
- For every 50 enemies in a row Hero destroys (indicated by a K.O. counter), he also earns an extra life—as long as he does not get hit by an enemy or do any damage to himself.

But alas, like all of us, Hero has his limits. After you watch a tutorial that teaches you how to control Hero, you engage in the game world. You successfully defeat a few enemies, and then you approach a bed of fire and think: "That sure doesn't look safe!"

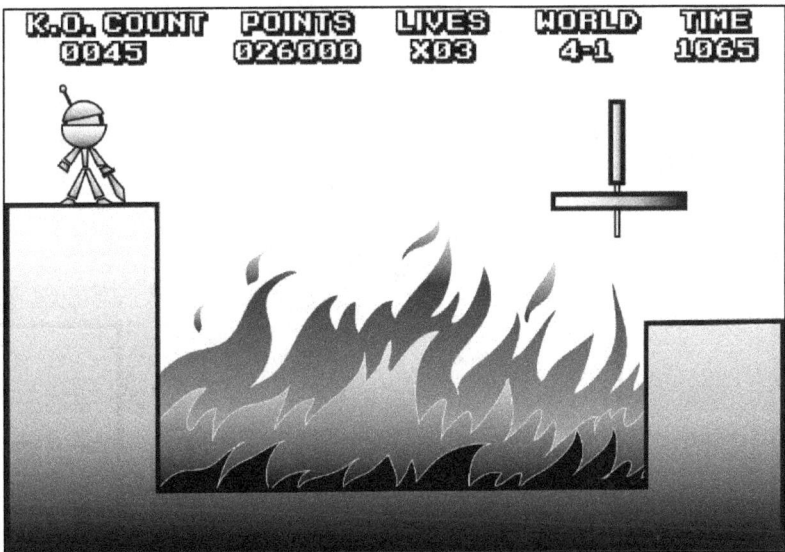

K.O. COUNT	POINTS	LIVES	WORLD	TIME
0045	026000	X03	4-1	1065

Being the daring gamer you are, you decide to jump in and see what happens. Maybe Hero's armor can withstand the fire?

Or maybe not...

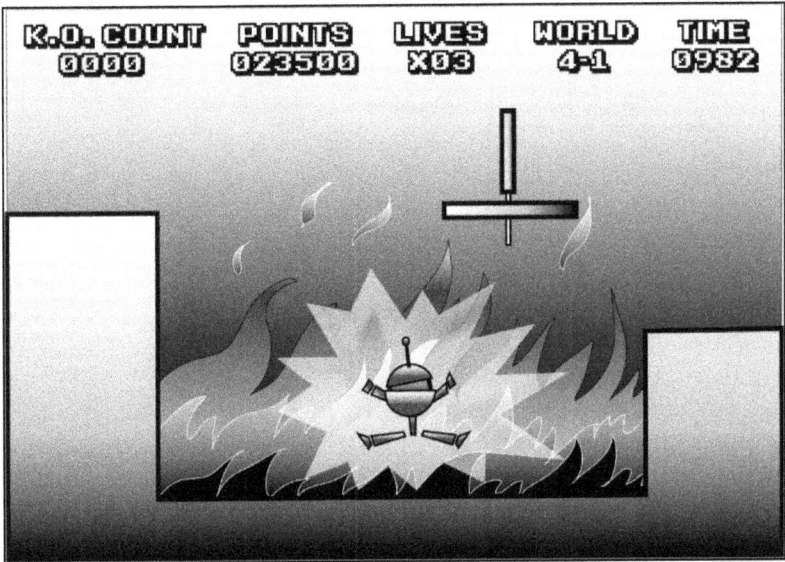

"ACK!"

Because you landed in the fire, Hero gags in pain and flails backwards in an uncontrollable but somewhat hilarious fashion. An ominous sound effect plays—it is similar to a teacher scratching a chalkboard! You are penalized by the game, losing points, and your K.O. counter resets to zero.

Oh and don't look now, but Hero is blinking as though he is about to fade away! Say what?!

YOU HAVE OFFICIALLY LEARNED THAT YOU SHOULD TRY
YOUR BEST NOT TO LAND IN FIRE IN THE GAME WORLD
OF *ADVENTURE GAME.*

Next time, you'll surely use Hero's jumping abilities and any
available platforms in the game world to avoid the fire,
won't you?

While this is a very (very!) basic example, we should not take the learning experience that is represented here for granted. A player can learn what to avoid in *Adventure Game* through the penalties that the game invokes, or by how certain choices prevent him or her from progressing toward the goals of the stage, level, or board. *Adventure Game* penalizes the player for a bad move (landing in fire) by taking away points and resetting the K.O. counter. The game also emphasizes how bad this move is by making it impossible for the player to control Hero immediately afterwards.

Additionally, the game uses other affordances of the video game medium to drive the point home. Hero's entire body flails backward and blinks when he lands in the fire; if you saw someone flail backwards, and his or her entire body started to blink, you would be quite startled, right? Also, a piercing sound effect plays when Hero lands in the fire. This sound is similar to when a teacher scratches a chalkboard. The sound of a teacher scratching a chalkboard is the epitome of unpleasant. Even *writing* about a teacher scratching a chalkboard is making me cringe! This disharmonious sound effect reinforces how bad of a move it was to jump into the fire. These disruptive visual and audio effects teach the player that he or she should avoid letting Hero land in fire – and likely other treacherous looking areas – in the future.

On the opposite side of the coin (yes, I'm using coin because of the whole video game thing), *Adventure Game* can try to help players understand what counts as good moves in the game world. Two of the methods mentioned in the example are the points Hero receives for each enemy he pounces on and the rewards that Hero earns at key milestones (10,000 points or 50 enemies destroyed in a row = extra life). Visual and audio cues can reinforce those positive developments, such as the game blasting a distinct, heroic musical tune whenever Hero destroys the 50th enemy in a row or earns the 10,000th point to receive an extra life.

All of these elements in the game world, along with in-game tutorials and any other helpful resources planted in the video game by the design team, can help players progress through the game. The role of the player is to navigate the cues of positive progression and negative setbacks to get Hero through *Adventure Game's* levels, bosses, sub-quests, and mini-missions. Just as there is an interdependent relationship between video game and player, there is a nearly symbiotic relationship between video games and the concept of learning. Players have to learn in order to progress through *Adventure Game* because *that's just how games are designed.*

The learning that can occur when playing a video game reaches far greater levels of complexity than what I demonstrate in the *Adventure Game* example. Players have to consistently learn and adapt to solve problems and progress in video games, whether they need to figure out how to execute new moves, understand the appropriate time to use new powers, interpret physical and verbal clues, or recognize dangerous patterns to avoid. All of this learning is baked right into video games, embedded to the point that players may not realize it (just like the learning that happens in our everyday lives). As I touched on earlier, the value of video games as learning environments is substantial.

It makes sense then, that a medium that fosters inherent problem solving and learning on such a prominent scale would be leveraged, if possible, for learning that can be transferred to real-life situations. Although the objectives of these two forms of learning differ (one is learning in order to get through a game, the other is learning academic or soft skill subject matter to not only get through a game, but to also get better at the skill in the real world), the fact that video games keep players engaged for continuous learning is a promising start. This makes them attractive for those who would like to create learning experiences that benefit players beyond the game. In Level 3, I will discuss games that have been created with this purpose in mind and some of the momentum these learning games have generated in recent years.

Level 3

Enter the Learning Game

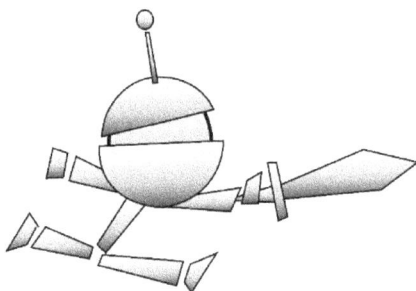

For decades, game designers and researchers have worked to create video games that marry the problem-solving intricacies of gaming with the complicated goal of facilitating knowledge that transfers to the real world. In learning games, performing a character's special move would serve a grander goal of learning how to, for example, solve algebraic equations. Solving those algebraic equations would be necessary to progress through the game. The ultimate hope, however, is that players will not only learn how to solve the equations within the game world, but will strengthen their ability to solve these sorts of equations outside of the game as well.

There have been notable successes in the learning game genre. Some of the most prominent titles include *The Oregon Trail, Where in the World is Carmen Sandiego?* and *Math Blaster. The Oregon Trail,* first developed in 1971 and recently inducted into the World Video Game Hall of Fame (The Strong National Museum of Play, 2016a) gives players the opportunity to "assume the role of Western settlers crossing the continent on the way to the Pacific Coast" (The Strong National Museum of Play, 2016b, para. 4). Set in 1848, the game simulates pioneer life and the decision-making that accompanies it. Accordingly, players have to make appropriate purchases to continue their journey, while remaining mindful of food supply and pending disease.

Because the game takes place in 1848, players "live" through and essentially learn American history in their roles as the Western settlers. Any conversation that talks about video games and learning usually pays respect to this game. The prevailing reality is that *The Oregon Trail* has married video gaming with knowledge facilitation (the simulation of traveling as a Western settler in 1848) in a massively successful way, selling over 65 million copies to date (The Strong National Museum of Play, 2016b, para. 4). The commercial and mainstream success of this learning game is (or at least, should be) a tremendous building block in advocacy efforts for learning games in general.

The same can be said for *Where in the World is Carmen Sandiego?* First published in 1985, this learning game series is one part of a larger educational multimedia empire that has included live-action television shows and cartoons (Houghton Mifflin Harcourt, 2015). The original game focused on helping players learn geography, prompting players to use clues to find where the thief Carmen Sandiego and her allies were located. The sprawling success of this series, and the iconic status it has achieved over time, provide another strong foundation for advocacy efforts for learning games.

My favorite of the three is *Math Blaster*. Originally developed in 1983 (The Internet Archive, n.d.), this game put players in the role of an astronaut who explores various intergalactic terrains while solving math problems. More than any other game from my younger years, *Math Blaster* provided me with a learning experience that I transferred to real-life contexts. I distinctly remember getting better at multiplication through this game, and this was entirely due to the fact that the game just kept throwing math problems at me! The *Math Blaster* series includes many games and has also become critical to any conversation on learning games.

These games of decades past were certainly advanced for their time, considering the technological limitations and shortage of empirical research on video games and learning. They helped set the foundation for the current era of games that are reshaping what "advanced" in a learning game looks like. This era of advanced learning games include titles with the type of intricate storylines, complex problems, and immersive gameplay that is possible with modern technology and our better understanding of the video games and learning dynamic.

The rise of digital publishing has opened the door for designers and developers to distribute learning games of all types through various channels. While titles from the past were mainly confined to desktop computers, the learning games of today can also be played on smartphones, tablets, and many other devices. They are as spontaneously accessible as they've ever been, with some just a mouse click or touchscreen button-push away. Advanced learning games share this benefit, meaning that many people now have instant access to powerful learning experiences that take significant advantage of the video game medium.

Some advanced learning games are characterized by a traditional platform experience that is synonymous with entertainment games. An example of this is *Treefrog Treasure*, a math game from the University of Washington's Center for Game Science. The game puts the player in control of a frog who must break chains by kicking a number, symbol, fraction, or space between these characters that matches what the game displays next to the chain. For example, the chain may have a number line that only says (from bottom to top) 0%, 30%, 60%, and 100%. The number the game may display next to the chain could be 90%. The player must aim the frog to kick the chain where 90% should be (right under the 100%). If the player is right, the chain will break, and several gems will appear that the frog can collect. If the player is wrong, the frog stops in his tracks the second he hits the chain and looks like he's been physically stunned.

Treefrog Treasure ups the difficulty as the player progresses through the colorful game world. As an added challenge, players have the option of collecting all of the gems in the level (which requires precise hopping with the frog). Players are also rated 1-3 stars after each level based on their performance. In creating a game meant to help players with math concepts, the Center for Game Science also crafted a complete video game experience with enough extras to give players incentive to replay the game. Like *Math Blaster*, the repetition of solving various math-related problems hopefully translates to increased capabilities in the real world for the player.

There are other advanced learning games characterized by gaming experiences that are more on the unconventional side. One example is *Decisions that Matter,* an interactive graphic novel-like game created by students at Carnegie Mellon University. This game addresses sexual assault on college campuses by putting players in control of a college student's various decisions throughout a day. The student, as a bystander, has a few opportunities to intervene when a female friend faces potentially uncomfortable advances from males. The player gets to choose if (and to an extent, how) the bystander should intervene. Ultimately, these decisions will decide the fate of the female friend, with some choices putting that female more at risk of sexual assault at a party later that night. While the game takes on the look and feel of a graphic novel, the ability to make choices about a student's decisions with variable consequences positions this firmly as a game.

While *Decisions that Matter* may not be considered a learning game in the same sense as *Math Blaster*, the game provides a learning experience that raises awareness of the implications of decision-making — even as a bystander — and the risk of sexual assault. In this regard, it is more akin to how *The Oregon Trail* simulates the endeavors of Western settlers, the conditions that they endured, and the decisions they had to make. The game itself aims to be a prevention resource for college students, which means that what is learned in the game is certainly meant to be transferred toward wise decision-making in the real world (especially on campus).

As we see in *Decisions that Matter,* advanced learning games in this current era are pushing boundaries, content-wise. In *Fair Play* by GLS Studios (based out of the University of Wisconsin-Madison), players learn about various implicit biases while walking in the shoes of a Black graduate student named Jamal. Jamal is trying to rise in the ranks of academia, and facing these biases (most of which are racially-charged) present real challenges in his journey. The goal of this game is for players to recognize the implicit biases that have so casually become a part of our everyday lives. I've used this game as part of a research methods course I teach to help my students recognize how biases can affect research. This game helped our class continuously think about how to avoid biased thinking. *Fair Play* turned a potentially disastrous subject matter into something that could be experienced and constructively analyzed.

Another advanced learning game that pushes boundaries content-wise is *Spent* from the Urban Ministries of Durham. This game asks the player to make decisions on how to spend money while living in poverty. *Spent* is a prime example of how advanced learning games do not need flashy graphics or boisterous background music; the majority of the game is text-based with minimal, 2-D graphics. Yet the game immerses the player into a life where every decision is absolutely critical. A dollar balance remains in the corner of the screen, dwindling after the player makes difficult financial choices and simultaneously increasing the pressure on the player throughout the game. This game puts players into the shoes of people who go through this type of stress every day. The design of the game has the capability of not only evoking empathy from the player, but also helping the player further realize the ramifications of their daily financial decisions in the real world.

What we see in this small sample of advanced learning games are the possibilities that are produced when we marry the whimsical nature of video gaming with the complexities of learning. We can create learning experiences to advance knowledge in a specific subject matter, or help people learn more about the challenges that people face in their everyday lives. Further, the video game medium allows designers to create these experiences in creative and engaging ways, and even tinker with the standards of what a video game is "expected" to look like.

The examples I list are just a mere morsel of the phenomenal advanced learning games that are being developed in the modern day. Such games were not technologically routine in the early days of learning games. As we have entered this new era, which consists of learning games that break previous gameplay boundaries, the conversation on the merit of video games for transferable learning has escalated. The 2016 National Educational Technology Plan from the U.S. Department of Educational Technology highlights a primary reason why:

> "At a higher level of engagement, digital tools such as games, websites, and digital books can be designed to meet the needs of a

range of learners, from novices to experts" (U.S. Department of Education, Office of Educational Technology, 2016, p. 19).

Ultimately, the most promising aspect of learning games in general, and advanced learning games in particular, is the possibility of reaching a wide array of learners through the power of a very popular medium.

Yet all of this is for naught if *people do not play these games*. More on this in a bit.

GIVING ADVANCED LEARNING GAMES A CHANCE

There has been a surge of passionate advocacy this decade that has explored the benefits of video games and learning. This, by default, benefits the very idea of the learning game, and advanced learning games by extension. One major effort in recent years originated from the White House. In 2011, Professor of Informatics Constance Steinkuehler (University of California, Irvine), a fervent researcher on the benefits of video games for learning, cognition, and social change, was appointed Senior Policy Analyst in the Office of Science and Technology Policy. Her role consisted of advising on policy related to video games and learning (Higher Education Video Game Alliance, 2016). This indicated an important realization from national officials that video games offer much to the world, and we have only begun to scratch the surface of their potential.

Likewise, the U.S. Department of Education has hosted several programs highlighting video games as learning resources. In 2014 the Department hosted Ed Games Week, a week of events focused around the discussion of the role educational video games should have in the classroom (Metz, Shilling, & DeLoura, 2014). As part of that week, the White House hosted an Education Game Jam, where thousands of game developers joined teachers, researchers, and students to build game prototypes focused on K-12 academic subjects (DeLoura, 2014).

A slew of online newspaper articles have kept the wheels going in the video games and learning conversation, with headlines such as, "How Video Games Can Make our Kids Smarter and Learning More Engaging" (Matthews, 2015). Also, there have been a steady number of noteworthy books that have explored the topic. Kurt Squire's book, *Video Games and Learning: Teaching and Participatory Culture in the Digital Age* (2011) is an extensive and carefully crafted work that finely details the nuances of video games and learning. A fairly recent work is education reporter Greg Toppo's *The Game Believes in You: How Digital Play Can Make Our Kids Smarter* (2015). This book has garnered praise for the insight Toppo provides on the way educators are using games for learning purposes.

This ongoing discussion on video games and learning is encouraging for the future of advanced learning games. Perhaps more critical to their future are the researchers who are working hard to determine whether advanced learning games are actually effective. The aforementioned Center for Game Science is one of several prominent research centers that have emerged in universities to study the dynamics of video games and learning. There are also non-academic organizations who are exploring the potential of learning games, such as GlassLab Games (Redwood City, CA) and the Joan Ganz Cooney Center (New York, NY).

Bearing this in mind, I believe that people are willing to give this video games and learning idea serious consideration due to the following factors:

1. the overwhelming popularity of the video game industry;

2. the inherent problem-solving dynamics of video games;

3. the rise of resources and institutions dedicated to exploring the learning benefits of video games; and

4. the ability to create lower-cost, experimental games thanks to advances in computer and digital technology.

While it is encouraging that more people are seriously discussing the benefits of the video game medium for learning purposes, the ultimate question remains: do games actually foster learning that can be transferred to real-life contexts? This brings us back to the stark reality that the video game medium is still relatively young, and the high level of attention for complex, transferable learning from video games is even younger. We simply cannot definitively say that advanced learning games are automatic pathways to transferable learning.

The fact that we cannot strongly say yes to the "do they work?" question is a major roadblock for the future of advanced learning games. This is especially true since *less advanced* learning games, which may have lower production costs and take less time to learn how to play, may be more feasible options for publishing and consumption. Also, there are other forms of media that have a much longer history of being trusted as learning resources, such as books. These factors make it less of an imperative for anyone to actually give advanced learning games much of a chance.

This is why I maintain that the video game is the learning technology of the *future*, not quite the present! Designers of advanced learning games are still trying to determine the best ways to situate transferable learning experiences within games. There is no consensus about the way to do this. Just because a player has to learn to get through a game does not mean that the player will take that knowledge with him or her to the real world.

Further, how should this potential learning be measured in the game and in real life? Although research centers that focus on (and advocate for) advanced learning games have reported some learning successes from these games, more third-party, scientific research and case studies are needed.

This means that there is much that still needs to be evaluated about the design and efficacy of advanced learning games. While there has definitely been a rising tide of positive rhetoric on video games as learning tools, the shortage of hard data to support this notion certainly warrants hesitation. As such, the big question is whether advanced learning game development is even worth our time, energy, and especially our dollars (as producers or consumers).

As contradictory as this may sound, this type of cautiousness is exactly what this learning technology needs. It is a sense of caution that many learning technologies have not had the benefit of receiving. By taking time to pause and ask the right questions, we can position advanced learning games to reach their full potential as powerful resources for learning in the future.

The key is for video games and learning advocates to keep up the momentum on this topic. Video games are so popular, and the medium is unlike any other, that we must seriously continue to explore if we can use these games for the positive outcome of transferable learning.

> **As educators, scholars, and researchers, this is just the type of challenge we prepare for, right?**

MY CONTRIBUTION

As someone who spends countless hours advising college students on how to navigate the world, I can't help but think of the larger, sociological picture of almost everything! So of course I think about the broader implications of advanced learning games at great length. While researchers and educators continue to produce data and case studies on this topic, I believe it's also important to analyze how advanced learning games fit into our society at large.

There are two key parts of a learning game in general — the video game itself, and the knowledge that is meant to be transferred from the game into the real world. The video game aspect is all about interactivity, play, and engagement, while the knowledge aspect is all about comprehension, understanding, and interpretation. The aspects of play and learning are not mutually exclusive in learning games, yet for various reasons our society often places these aspects at odds with each other. If society does not encourage the synergy between play and learning that a learning game attempts to produce, is it realistic to expect advanced learning games to succeed in that society? Is it realistic to expect people to get excited about playing and perhaps purchasing these games?

To answer these questions, it is necessary to explore the problems that result from the incompatibilities between learning games and society at-large. The next four levels will detail some of these problems to get the conversation started, but there are certainly many more issues than what I cover in these pages. While it is important to research design and assessment in advanced learning games, if these games cannot find a successful spot in mainstream society beyond the classroom, then that is a wasted opportunity.

We have the chance to create truly transformational learning experiences using a medium that can take players to worlds unknown. Yet it is not worth developing these games with such depth if not enough people will play them. Exploring the two paradoxes and the two tensions that follow, and eventually addressing them, could go a long way in attracting more players to advanced learning games and securing a brighter future for this form of learning technology.

Level 4

The Video Game Paradox

A recent Entertainment Software Association report on the computer and video game industry notes that 63% of U.S. households are home to at least one person who plays video games regularly (Entertainment Software Association, 2016). To give a demographic example, the Pew Research Center found that nearly "72% of teens play video games online or on their phone" (Lenhart, 2015). That is a massive amount of people! If all of these people play video games, surely that is a large potential audience for advanced learning games, right?

It's not that simple. One of the biggest challenges for advanced learning games is the fact that they are video games. This may sound like a reversal from the rhetoric of the past two levels, but such is the case. While many people have discussed the video game medium's potential for facilitating transferable learning experiences, the technology known as the "video game" has become synonymous with the idea of entertainment. This automatically creates a paradox, where the very thing that is championed as great for advanced learning games (being a video game) can also be detrimental to their success.

Entertainment games are usually what we think of first, second, and third when we think of video games. Consequently, video games made for entertainment purposes are positioned to benefit the most as the industry continues to grow. It is no coincidence that the top 20 selling video games of 2015 were all entertainment games such as *Call of Duty: Black Ops III*, *Madden NFL 16*, and *Fallout 4* (Entertainment Software Association, 2016). The list of the top 20 selling computer games were all entertainment games as well, including titles such as *Elder Scrolls*

V: Skyrim and *The Sims* 4 (Entertainment Software Association, 2016).

> ## Unfortunately, there is no "Really Cool Game that Teaches Emotional Intelligence!" to be found on the list.

Since entertainment games have experienced substantial commercial success, they are the ones that get widely produced. This production creates a marketplace where entertainment games dominate other types of video games. If entertainment games dominate the marketplace, they stand a high chance of also flooding the eyes and ears of potential video game players. Here is where we have to consider the old principle of supply and demand, or at least the foundation of the principle. There is a higher demand for entertainment games than other types. Designers and publishers, who need to earn a living, oblige to this demand by developing and publishing these type of games.

Players see entertainment games the most and may know others who have entertainment games. These type of games fill physical and digital shelves, are the main subjects of web and television marketing campaigns, and are the topics of most media coverage that video games receive. This makes purchasing an entertainment game an attractive option for remaining connected to mainstream society. This also leaves very little room for the mainstream focus on video games to be on anything besides entertainment-related titles. All of these factors combine to reinforce the connotation that video games are entertainment games, and anything else has to sort of prove itself worthy of the video game label.

Entertainment games sustain this prominence in the marketplace, entrenched into everyone's minds as the definitive representation for video games. The cycle repeats: these video games get the time and attention of potential players, the dollars invested into their purchase, and the funding for their design and publishing. If all of these resources primarily go to entertainment games, how much is left for video games that do not fall under that category?

All of this presents us with a deep-rooted challenge for advanced learning games: is there an audience for these games outside of education? These learning games attempt to take significant advantage of the video game medium, and by doing so will draw comparison to entertainment games that are already expected to do likewise. While this opens up advanced learning games to a large potential audience as outlined earlier in this level, it directly sets them up under the mounds of competition within the entire video game industry. It's tough enough for any single entertainment game to do well amidst this competition, let alone a game that tries to tweak that entertainment formula.

Video games and learning advocates have to think about how to motivate people to produce advanced learning games when the demand suggests that entertainment games would be a wiser time and financial investment. Of course there is the altruistic motivation based on the social good that could come from producing advanced learning games, but I believe that can only go so far. How do we incentivize designers and publishers to spend their time and money creating great games with transferable learning experiences? Even if we can encourage more people toward advanced learning game development, are there enough resources available for the development of these games, given that entertainment games will likely always receive priority?

REVERING THE ADVANCED LEARNING GAME

Perhaps the following is true: if we can compel more people to play advanced learning games with a comparable level of fanaticism as they do entertainment games, we may see a future where advanced learning games receive increased attention and priority under the video game umbrella. If we can create this reality, or come close to it, designers and publishers of advanced learning games would ideally receive the resources they need to feverishly produce these games. The demand would necessitate the supply. The result would be a surge of games that tackle learning in all areas, from math and science, to overcoming personal adversity such as being bullied at school or even in the workplace.

A rise in the prominence of advanced learning games, paired with the popularity of video games as a whole, could lead to even greater success for the entire video game industry. There are too many times where we pit learning games vs. entertainment games in an either-or discussion. Maybe the "Game that Teaches You Emotional Intelligence" could join the top 20 selling video games list, but only because that game added to the overall sales and popularity of all video games produced that year (as opposed to taking away sales from the entertainment games).

Even noncommercial advanced learning games, like the freely available games I mention in Level 3 and games produced by nonprofits, universities, and research centers would benefit. These games typically rely on grants and donations for development and distributive purposes. If there is a high demand in society at-large for advanced learning games, that should provide a catalyst for noncommercial titles to receive funding for development from grant-bearing entities and donors.

Maybe this vision seems far-fetched, but we have seen other media used adequately for learning and entertainment purposes while sustaining demand. For example, books come in a variety of forms, with some geared toward entertaining audiences, and many geared toward helping people learn a skill. There is a demand for these type of books, from self-help to how-to's, that necessitates their production on an annual basis. Also, consider film; we love our blockbusters in all of their box-office smashing glory. Yet we also consume documentaries, which can help us learn about topics in visual ways that cannot be done with most other media.

Advanced learning games can enjoy a success similar to informative books and documentaries. They would not replace entertainment games, but rather join them as legitimate options for the millions of people who play video games every day.

This leaves us with the daring thought: why shouldn't we try to produce and revere advanced learning games on a scale comparable to entertainment games?

A key to reaching this future is for people to choose to play advanced learning games on their own time, not intricately tied to a school assignment, homework, or training. Players would make the choice to play advanced learning games because of a genuine excitement to play the games and further their learning in certain subject areas and topics. Players may even choose the advanced learning game over the entertainment game at times, much like a person may choose to watch a documentary over watching the latest blockbuster entertainment movie. Either way, advanced learning games would be viewed as legitimate, daily options for game players.

BEYOND THE PARADOX

How do we reach a future where advanced learning games are held in a comparable regard to entertainment games, canceling the video game paradox? How can we get to a future where we see advanced learning games in the top 20 selling video and computer games lists? These are important and tricky questions to address. We do not want advanced learning games to be so much like entertainment games that they lose the identity and character that should compose a learning game. These games should ultimately focus on helping players learn some type of subject matter, topic, skill, or way of life in a manner that has the possibility of transferring over to the real world. Can a game achieve such a lofty goal and succeed in a world full of video game behemoths like the *Tekken* and *Final Fantasy* franchises?

The answer may lie in the interdependent relationship between video game and player that I describe in Level 1. That relationship has buoyed the popularity of the video game industry. Video games mean a lot to our society and are highly personal to us because these games need us. Yet only the entertainment game has *fully* reaped the benefits of this relationship. Perhaps there are ways that advanced learning games can develop similarly strong—or maybe even stronger—relationships with players everywhere.

Whether this is true or not, it's important for video games and learning advocates to bear in mind the following: the very thing that excites people about the video games and learning dynamic (the video game medium) is also something that holds advanced learning games back. This is due to the medium's close association with entertainment games. The only way to address this paradox is to acknowledge it, embrace the realities of it, and work towards a future where advanced learning games coexist with entertainment games to spur the video game industry as a whole.

Level 5

The Learning Paradox

The video game paradox is one obstacle to the future success of advanced learning games, but perhaps a bigger challenge is what I will classify as the learning paradox. In this case, the primary purpose of an advanced learning game—facilitating transferable learning experiences—also serves as self-defeating due to our society's perspective on the *idea* of learning. This perspective, in my view, associates learning with the hardships of work rather than the cathartic nature of play. I believe the learning paradox is worse than the video game paradox because it weakens the very concept that underlies the existence of learning games!

Maybe I'm off track here, but it seems that our society has a love/hate relationship with the concept of learning. We love what learning represents and the positive outcomes that are produced when we learn something. We consider education as one of the great equalizers in our society because of what every individual can learn within this institution. We praise intelligence and encourage the pursuit of personal knowledge through books, courses, training, and a host of other avenues.

However, in my lifetime I feel that the idea of learning has been positioned as diametrically opposed to the idea of fun or play, as if these concepts do not naturally work hand-in-hand. The idea of learning in our society is predominantly linked to non-entertainment capacities such as schooling and training. In these institutions, work is usually more of a priority than play. As a result, we may make the correlation that learning equals work, and that learning and play are accordingly on opposite ends of a spectrum.

As a result, the idea of learning can sometimes have an unfavorable connotation (there's that word again), more analogous to the sacrifices associated with work as opposed to the feelings of delight and joy that may be associated with play. The irony of this analogy is that it creates a dichotomy that is fundamentally incorrect. Learning is an entrenched part of play! It is a constant within play, whether we are learning how to adapt to the rules, or learning a strategy to succeed within the confines of an activity.

In my 2014 interview for *The Huffington Post* with Kevin Carroll (author, social change agent, and a fervent advocate for the positive benefits of play), he emphasized the connection between play and learning:

> *"But playtime was also productive time, even if as kids we did not realize it. What we thought was entertaining was also instructive. Activities we called tea party, kick-ball, finger-painting, hide-and-seek, daydreaming, and tag were also exercises in planning, strategy, design, decision-making, creativity, and risk-taking"(Wright, 2014a, para. 6).*

Carroll describes an ongoing dynamic of personal development that happens during play. This development is ignited by the opportunity for the player to interpret instruction and grow in skills such as design and decision-making. Accordingly, playtime also equals a time to spontaneously learn what it takes to engage in that session of play. We are usually just so overwhelmed by the entertainment aspect of play that we may not notice that learning is actually taking place.

We might also fail to notice because of our society's view of playtime as something that happens during designated periods *beyond* times of learning (e.g., recess in schools) or anything else that's not considered fun (e.g., an employee softball game to build camaraderie and break up the routine of office work). This social perspective of playtime, combined with the prevailing connotations of learning and play, add to the distance between the concept of learning and the concept of play in mainstream society. The resulting dichotomy produces an ongoing ideological tension between the two concepts.

Learning games, unfortunately, have to deal with the learning vs. play conundrum that is produced by this dichotomy. This puts advanced learning games in a precarious position. On the one hand, they are video games, and we play video games because they promise to be fun, right? On the other hand, learning is purposely the goal of these games, which may unfairly prompt the question:

> ## "Is this learning going to ruin my fun?"

Or worse, the thought that...

> ## "This learning <u>better</u> not ruin my fun!"

ACKNOWLEDGING THE LEARNING

The unfavorable connotations that learning has received has convinced some people that it might be worth "tricking" players to learn in learning games. This means that transferable learning moments would be so covert within a game that players would learn the subject matter or topic without recognizing it. The games themselves may not even be called learning games (or educational games, or any other similar label). This way, people would play the games, have fun, and learn along the way without being repelled by the idea that they are actually learning something.

I am not a fan of this mindset at all; it makes it seem like *learning* is "the word that must not be spoken!"

Beyond that, the idea of metacognition is important to me, considering my background in the learning sciences. Metacognition is when we actively think about how we are learning as we are learning. For example, if I am reading a book on repairing cars, I may actively think about various factors that affect how well I will learn from that book (e.g., my reading ability, the time of day and where I am reading the book, the formatting of the book, etc.). With this in mind, I can adjust my reading approach or the learning environment to increase the likelihood that I will actually learn from the book. For instance, if I am reading the electronic version of the book on a tablet, and the font is really small, I may think to myself "I learn better when reading big fonts because bigger fonts help me concentrate." Subsequently, I could increase the font size to better my chances at concentrating and learning the material.

I believe that this is one of the most important aspects to learning and intellectual development. Through metacognition, our learning becomes the subject of our own critical analysis. By doing this, we can determine how to position ourselves to better learn in our current situations and in future situations as well.

In essence, we learn how to learn.

Notice how, in metacognition, there needs to be an active awareness of our learning in order to further develop ourselves as learners? If the learning that is supposed to happen in a game is deeply hidden to the point that it is unrecognizable, how can we best reflect on that learning? Maybe we would learn the subject matter or topic, but there is a chance we would not understand why or how we learned, nor what the learning even means for our future learning. By making the learning too covert in a game, we potentially decrease the overall learning that a player could develop from that game.

This is a major reason why I do not think we should shy away from calling these games what they are—learning games (or educational games, etc.). To shy away from this label detracts from a core strength of the game—the wonders of learning. We should embrace the label and celebrate that these games explore the complexities of learning, or else what is the point of even trying to use the video game medium for such purposes? Advanced learning games do not necessarily need to announce, "This is a learning experience!" throughout the game, but I'm not sure that a, "This is a game; ewww learning!" approach (or anti-learning approach, to be more proper) is helpful either.

BEYOND THE PARADOX

As we shape the future of advanced learning games, it is necessary to remain aware of the connotation of learning in our society, and how that connotation can serve as a self-defeating obstacle for these games. The idea of "learning as work" has reigned supreme over the reality that learning is an intricate part of play. As a result, advocating for anything that attempts to marry learning and play can feel like an exercise in contradiction.

This may change for the better, however, due to advances in technology. Information and knowledge are only seconds away thanks to high-speed internet access and our plethora of computer and mobile devices. As a result, the practice of "learning" is breaking through institutional boundaries and pervading into the minutiae of our everyday lives. This enables instances of learning that can happen at any stage in life (lifelong learning) and can happen anytime and anywhere.

These phenomena of lifelong and anytime/anywhere learning often require people to *actively choose* to participate in learning experiences outside of the confines of institutional requirements. These are not new practices, but modern technology has accelerated their relevance in our society. As lifelong and anytime/anywhere learning become more mainstream, and the idea of learning becomes less rigidly associated with institutions known for their work, it will be interesting to see if the connotation of learning shifts. Will the concept of learning become less aligned with the idea of work and more synonymous with the idea of play? If this were to happen, it would be a significant shift that positions advanced learning games as resources for fun *because* of the learning they bring about—not in spite of that learning.

Level 6

The Tension of Failure

The two paradoxes described in the previous levels turn what should be strengths of advanced learning games into potential weaknesses. There are other challenges to the future of advanced learning games that are not quite paradoxical, but produce resistance towards their "fit" in mainstream society. These tensions are fueled by social trends and realities, and require just as much analysis as the paradoxes. I will use this level and the next to discuss two tensions, although I am sure that this is only scratching the surface. The first is what I will refer to as the tension of failure.

There are many factors that can lead to failure in a video game. One example is when a player tries out a new method to overcome an obstacle in the game. Player 1 may have an inclination that a new strategy can get him or her past the problem. Player 1 then tests out that new strategy, and along the way, his or her character dies. Player 1 may still feel that this strategy could work, so he or she tries again. And again. And again. If Player 1 does not give up, he or she may eventually find a way for this new strategy to work, or might just use a different strategy.

This example depicts a vital feature of video games: that players can try new tactics in an environment with limited real-life consequences. This affordance is crucial for video game designers, who can accordingly create problems in a game that challenge — but hopefully not completely stump — players. The U.S. Department of Educational Technology highlights this dynamic in their 2015 *Ed Tech Developer's Guide*:

> *"Game designers are particularly adept at motivating persistence, and much can be learned from the methods they use to inspire players to persevere in the face of difficulty and frustration" (U.S.*

Department of Education, Office of Educational Technology, 2015, p. 10).

Game designers place challenges of various difficulties in games, and are simultaneously tasked with creating an experience that encourages players to overcome those challenges in the face of failure. In turn, players can use various strategies to engage these challenges. If Player 1 uses a strategy and it leads to a negative consequence (losing all hit points or gold coins, for example) he or she can choose to tweak the strategy or use a new one altogether to pursue better results. Player 1's option to restart, reset, or retry the game creates an environment where the player is free to fail within the game world. Because the consequence of failure in video games is tied primarily to the game world and not the real world, Player 1 can use failure as a tool to learn how to play the game better.

Game designer Lindsey Tropf (Founder and CEO of Immersed Games) stressed this sentiment in my 2014 interview with her for *The Huffington Post* regarding the benefits of game-based learning:

"Risk is minimized because you can build a tower without it collapsing, or perform a surgery without harming a patient. Students are free to experiment and learn from failures" (Wright, 2014b, para. 5).

Ms. Tropf highlights the range of possibilities available to video game players. Players can virtually try things without worrying about the negative consequences that could happen if they tried to do those same things in real life. These attempts take place within a fictional world, which means that a player can fail as many times as necessary to learn how to get past a problem. In the midst of this failure, the player may learn new methods and approaches that can help him or her progress through the game.

This is not to minimize the unpleasant feelings that failure in a video game can cause. Players do what they can to fend off failure in video games to avoid the frustration of not completing tasks, missions, goals, levels, and the game itself. In a multiplayer game, Player 1's failure could be Player 2's path to bragging. Who wants to be on the wrong end of that? If multiple people are waiting for their turn to play a video game, failure may mean that a player has to give up the controller and even wait a long time to play again. Also, video games can go over-the-top in shaming a player for falling short of game goals. For example, some characters in these games literally fall off of the screen every time they die. Seriously, where do these characters go?

Failure in video games can indeed be a disastrous moment for players, causing everything to grind to a halt or forcing a player to face social scrutiny from other players and observers. Even with all of that in mind, failing is an entrenched and necessary component of the relationship between video game and player. Without the presence and reality of failure, how can players truly get better in games, solve problems in new ways, and develop their gaming styles?

FAILURE IN THE REAL WORLD

There is a certain appreciation necessary for the role that failure has in video games, even if we do not specifically label every setback or defeat in these games under the category of "failure." This appreciation does not seem to fully extend at all to society in general. While I believe that the same concept of "failure leading to learning" holds true in the real world, the stakes of failing in our society are often too high for this concept to have much value. As a result, failing is often treated as the antithesis of learning, something to be wholly avoided, rather than a valuable component of learning. This especially applies in the institution of education, which may inadvertently set the foundation for this type of thinking throughout our entire lives.

For example, an "F" on a math test in school is typically regarded as a devastating occurrence for students of all ages. Unlike failure in video games, there is not always an easy way to restart, reset or retry before panic sets in for the student. In literal terms, an F on a test means that, for whatever reason, the student did not earn enough points to receive a passing grade. This could have happened due to many reasons; perhaps the student did not get enough sleep the night before, or missed breakfast and was too hungry to concentrate.

Nevertheless, in our society, we swiftly attach negative attributes to an F regardless of why it happened. The school may unfortunately apply these negative attributes to the student as a whole, so that a failure on a math test quickly leads to, "This student does not know math." As a result, the student may be handled in ways that are considered appropriate for someone who "does not know math," such as being assigned remedial work.

In this example, failure is stigmatic and something to be fixed or corrected. This can foster an environment where students only want to know what it takes to *not* fail, as opposed to feeling motivated to discover the different ways they can overcome challenges. For some, experimenting with different ways to solve a problem is simply not worth the consequences that failing to correctly solve that problem may bring. Thus, tried-and-true methods of getting through problems become highly attractive options, along with even unethical methods such as cheating and plagiarism.

This beckons the question:

> **If we stigmatize failure as early as grade school, how can we expect to cultivate a society that encourages failure that could lead to learning?**

I'm not sure that this is possible. As we move on from school to the different phases of adulthood, our responsibilities grow along with the stakes of our daily choices. Much like in the game *Spent*, every decision we make as adults has reverberating implications, whether it's what career to pursue, when to buy a house, or what health insurance plan we should select. If we make a mistake of some sort during these decisions, there's usually not much time to bask in the glow of what can be learned from that mistake due to our fast-paced, competitive society. In essence, failure *can* be a learning tool, but socially it is generally considered more of a debilitating nuisance.

This is where the tension lies; we must be careful that the social stigma we have attached to failure does not cross over to advanced learning gameplay. As I have stressed, video games are safe places to fail, and failing is an essential part of these games, but they are juxtaposed with a society that primarily discourages failing from an early age. The stigma of failure in our society is automatically at odds with what is possible and encouraged in a video game.

Once we make transferable learning the primary goal of a video game, as it is in schools, does the stigma of failure cancel out some of the freedom to fail within the game? Will players be apt to show that they can "get the answers right" in the game as opposed to exploring new strategies and methods to learn and overcome problems in the game? Will players only want to show they can get the answers right without really understanding (or caring) *how* they got them right? Will cheating take priority over learning in these games so that players can avoid failure? If any of this is the case, then there is no need for complex learning games that immerse players in engaging game worlds. In fact, there wouldn't really be a need for learning games at all; we might as well just give everyone flash cards.

BEYOND THE TENSION

There seems to be greater awareness today about the value of failure to the learning process. Phrases such as "failing forward" and "failing fast" show that people want to turn something that could be considered a negative (failure) into a positive tool for progress. The increasing popularity of processes that encourage swift idea development, testing, and revision—such as design thinking—is a clear indicator that people are recognizing the benefits of failure and the growth that can result from it.

However, our society cherishes accomplishment and achievement on too grand of a scale for there to be a quick, sweeping change in the attitude towards failure. Because of this, the stigma attached to failure will stick around for quite some time, and people will surely try to avoid failure—regardless of topic or setting—at all costs.

Somehow, video games and learning advocates must determine how to keep this stigma from enticing players to limit their creative approaches in advanced learning games. This may mean that these games have to be carefully designed so that failure equates to progress in the game, instead of failure equating to an end-of-game experience like in entertainment games. Also, those who use advanced learning games as part of their instructional activities can construct learning environments around the game that focus on the process of *getting to* the learning goals as much as actually achieving those goals. Whatever methods we try to use, it is important that we counter the tension of failure to maximize the affordances of the video game medium for learning.

Level 7

The Tension of Educational Gaming

The idea of *learning*, and what it represents, gets a bad reputation at times. Learning is such a complex phenomenon, one that requires thorough analysis and questioning to fully understand. However, as I mention in Level 1, we just do not have time for analysis and questioning! We want answers, we want them now, and we want to know what the next great thing is immediately.

The idea of learning falls victim to this need for immediacy. We love the potential rewards of applying learning goals to many things in society. For example, there are always professional development training opportunities, educational programs connected to museum exhibits, and camps (sports or otherwise) where people are supposed to enhance and learn certain skills. We create these learning environments because that just seems to be a great way to help individuals grow and develop.

However, we do not always dispense the time and effort necessary to ensure that these learning environments actually work! To illustrate: a program may be a successful learning experience for some participants, but not necessarily all of them. One thorough way to account for this discrepancy in future offerings of the program could be to:

- scientifically analyze the learning environment where the program takes place;
- conduct in-depth interviews with every current participant, learning more about their backgrounds while asking them what worked and what didn't work for them in the learning environment;
- gather information on the expectations and backgrounds of the next group of participants; and
- use a synthesis of all of this information to influence the design of the program's next offering.

Sounds like a lot of work, right?

This type of analysis would require many resources, including the oh-so-precious commodity of time. So it may be tempting for the creator of the program to leverage more immediate and less intensive methods to gain feedback on the program. For instance, the host of a professional development course may issue a survey at the end that asks participants a few questions about what they learned and what they wished they learned. The host can then use the results of the survey as a guide when modifying the next offering of that course. This method *might* help increase the effectiveness of the course going forward.

Since the survey would take much less time and effort than the extensive method I described earlier, I consider that survey to be a *shortcut* to a better learning environment. Will the survey help the host create as effective a learning environment as a longer, more rigorous process? That is to be debated. What is true, however, is that the shortcut allows the host to limit the amount of time (and ultimately money) invested into creating a better learning environment. This reduces some of the risks of offering the course in the future.

Bearing all of this in mind, advanced learning games face a reality characterized by the following:

1. We love the idea of learning.
2. Our society is strapped for time.
3. We are waiting for more hard data to help us create effective advanced learning games.
4. We need answers for learning *now*.

Contrary to the immediate expectations within our society, it will take quite some time for educators, researchers and designers to determine how to create the most effective advanced learning games on a consistent basis. This leads us to ask, is there a way *right now* that we can take advantage of the video game medium for transferable learning? Is there a shortcut?

Of course there is, and this is where we have to discuss the tension of educational gaming.

RISE OF THE EDUCATIONAL VERSIONS

The conversation is growing on the benefits of creating educational versions of entertainment games. Instead of creating advanced learning games completely from scratch, designers can add learning goals and lessons to entertainment games on the backend. These games can be an appealing option to educators who want to leverage the popularity of the video game medium to foster transferable learning experiences. They can be attractive to individuals who otherwise would not give learning games a chance. They can also appeal to designers and publishers who are in favor of the concept of video games and learning; since the game world's major framework is already in place in the entertainment version of the game, there is a foundation set for the design team to build on.

Yes, this is a bit of a shortcut to achieving transferable learning in video games, but that's not a bad thing. One example is the phenomenon of *Minecraft*. Created by game developer Markus "Notch" Persson, this open sandbox video game allows players to create and build pretty much whatever they want in order to survive. *Minecraft* has been tremendously successful as a commercial entertainment game, but has also garnered much praise for its capability to foster subject-matter learning. Players and educators have found a variety of ways to create learning experiences within the open world (the benefit of players being able to create anything they want) with the expectation that the learning can transfer to the real world. As a result, game developer studio Mojang released *Minecraft: Education Edition* in 2016, with new features tailored specifically for use in educational institutions.

Here we have an entertainment game repositioned as an educational—and in turn, learning—resource. The enduring popularity of *Minecraft* will likely go a long way towards the ultimate success of the education edition. Does this example mean that the answer to all of the challenges advanced learning games face is for designers to create educational versions of popular entertainment games?

DRAWBACK OF THE EDUCATIONAL VERSIONS

I believe that a problem with creating educational versions of entertainment games is that we lose the opportunity to see the full potential of advanced learning games that are built from scratch. In particular, we lose some of the benefits of backwards design, an instructional design method where the learning environment and measures of learning are created *after* an instructor determines the learning goals. In other words, everything can be specifically geared to what the instructor actually wants his or her learners to learn.

Advanced learning games that are built from scratch can be created with a backwards design mentality. The design team can determine the game's learning goals first and then design the game according to those goals. By creating advanced learning games this way, design teams can push the boundaries of their creativity while creating a virtual world that leads players to specific learning goals. Conceptually, there are very few if any restrictions in place before the design team starts to create the game. If the design team feels that a sports game is the best way to teach an English subject, they are free to create that particular game world around that learning goal.

Conversely, educational versions of entertainment games are married to the structure and characteristics of the original game. By default, these games are characterized by an entertainment foundation that is supplemented with learning goals. This means that the idea of transferable learning cannot be the primary focus of educational versions, because these games make their claim to fame *based on their entertainment foundation*. The very root of the game cannot be tailored for learning, because the root is already tailored for entertainment. The design team has to work within these confines as they add explicit learning goals to the game.

No matter how well the learning goals are bonded to the entertainment foundation, there will be a difference between what the game is able to accomplish for transferable learning, and what it could have accomplished if it was built from scratch with those learning goals in mind. The thing to debate is whether the difference is a significant or insignificant factor in our efforts to facilitate transferable learning to game players.

Even if the difference is an insignificant factor to our efforts, there is another potentially negative consequence of championing educational versions of entertainment games. Whether we realize it or not, educational versions of entertainment games can contribute to the idea that learning itself is not a good enough concept—for whatever reasons—to successfully carry a video game on its own. Instead, this reinforces the idea that players are drawn by entertainment brands (beckoning back to the whole video games are entertainment games connotation), and that the idea of learning should be *tacked on* to games under these brands.

Again, this is not a bad thing. At the end of the day, the more products that facilitate transferable learning experiences, the better! However, we must be careful not to fully migrate to the "grass is greener" task of adding learning content to entertainment games, and instead also remain committed to the arduous task of creating advanced learning games from the ground up. To leave behind this sort of commitment shortchanges gamers and learners of all ages, I believe. The complexities of learning warrant games that are built from the start with those complexities in mind. Ultimately, this can allow us to bring about one of the greatest possibilities of the video games and learning dynamic: games characterized by unimaginable learning experiences that are just not possible in games originally built for other purposes.

Endgame

The affordances of the video game medium have piqued the interest of many people—gamers, educators, researchers, even government officials—to consider how video games can be used for purposes beyond entertainment. Most prominently, there has been a wave of momentum in favor of video games as resources for transferable learning. The inherent problem-solving dynamics of video games have naturally extended our curiosity in this regard.

The momentum that advanced learning games have today is due in part to the early efforts of game designers who attempted to put the video games and learning combination into practice. Those designers created learning games such as *The Oregon Trail* without the technological advances of this century or the growing amount of literature and analysis on video games and learning that we have access to today. There should be expectation for the learning games of today to harbor more comprehensive, intricate learning experiences that are also engaging gaming experiences for players. The result is a new era of advanced learning games such as *Fair Play* that are challenging the traditional convention of how to facilitate transferable learning within a game world.

This new era of advanced learning games has been met with unbridled excitement at times, as well as appropriate waves of caution. There are a number of people who are eagerly trying to figure out the best way to design these games to host learning experiences for players that can be transferred to real-life situations. Despite this, we still do not have enough evidence that advanced learning games are truly worth our time, energy, and dollars. Unfortunately, we have to call it like it is: until we have more supportive data that advanced learning games consistently work for human learning, the video games and learning dream that we laud will continue to be an unfulfilled vision.

Still, it is important for video games and learning advocates to avoid tunnel-vision as we move toward the future of advanced learning games. Certainly, there should always be an imperative to focus on what design choices aid transferable learning in a game, and what in-game methods should be used to measure that learning. Yet there is an entire social world that we hope to integrate these advanced learning games into. This means that there will always be a social context to consider when analyzing these games. If we do not keep this in mind, efforts to create great learning games that truly impact people in society will be in vain.

I believe that these social contexts typically pit the concepts of learning and play against each other. This is in direct contrast with advanced learning games, as these games attempt to marry the complexities of learning with the wonders of video gaming. This contrast produces paradoxes and tensions that weaken the "fit" of advanced learning games in mainstream society. It is important to recognize these paradoxes and tensions, and in some ways we need to embrace their existence. Only then can we work *through* them and truly create the brightest future for advanced learning games.

Designing great advanced learning games is a wonderful feat, and overcoming the paradoxes and tensions that could stunt their success would be outstanding. However, we still need potential players to *know* about these games and desire to play them. Video games are incomplete without players, and advanced learning games are no exception. A major challenge for video games and learning advocates is to figure out what can compel people to play the fantastic advanced learning games that are being published today, and will be published in the future. In order for the future of advanced learning games to be brightest, players must feel compelled to actually choose to play these games aside from educational assignments or work training. This is as important an issue as the actual design of the games, the measures of assessment within them, and the integration of these games into learning environments.

> **If very few people play advanced learning games other than when they are assigned to do so, we have missed our opportunity.**

By moving away from the hyperbolic phase of video games as learning technologies, and facing some of the deep-rooted social challenges to their success head-on, we can create an important foundation for advanced learning games. This is the hope: a future where these games serve as important learning resources in and out of the classroom, while effectively attracting the eyes and ears of players of all ages.

BEYOND THIS BOOK

The major concepts I discuss in this book—video games, learning, play, work, and society—are not neat and tidy concepts at all. It's hard to talk about one without talking about another one in the group. This is why, in order to get to the brightest future for advanced learning games, it will take the efforts of many stakeholders. It will take collaborations across sectors and disciplines. Gamers, potential players, designers of entertainment and learning games, educators, and researchers should all be in the conversation. We also need to hear from those who may not be as interested in video games, but can bring valuable perspective from fields such as Media Studies, Sociology, and other pertinent social sciences. Yes, this is a huge melting pot of minds and perspectives that could be messy, but is absolutely necessary.

It can be argued that this kind of collaborative approach for *anything* would lead to escalated success, and if it was easy, it would happen more often. I know that collaboration across sectors and disciplines is actually very unwelcomed at times. However, we have in our midst a potentially groundbreaking technology for learning. If we settle only for learning games that marginally take advantage of the video game medium, we are missing a great opportunity. If we settle for advanced learning games that only reach a fraction of the audience that they could, we are shortchanging the many people who could benefit from these games. By including a wide array of voices in the conversation on the development and implementation of advanced learning games, we can give these games the best chance to succeed in a society that does not necessarily favor them. I do not think we will ever truly know how powerful video game technology can be for transferable learning (or if we should really just stop trying to use it for this purpose) without the appropriate collaboration of players and minds.

Let us not forget to also focus on the innovation aspect of it all, which itself deserves its own book and analysis. Augmented reality, improved virtual reality, and other immersive technologies are expanding the scope of what video games can look like in the 21st century. We must also remain attentive to game design approaches that incorporate multiple forms of technology and media (e.g., transmedia, mixed media, and alternate reality games). The affordances of these approaches may provide more flexibility for designers to create gaming experiences that truly foster transferable learning. It is quite possible that the brightest future for advanced learning games will need to include titles that leverage a combination of platforms. Perhaps an advanced learning game will need to start on a smartwatch, transition to an augmented reality application, and then finish up on a mixed-media or virtual reality platform. Innovations in technology have brought us to our current age of advanced learning games, and future innovations will hopefully produce games that consist of even greater transferable learning experiences.

As for me, I will continue to be a casual gamer, a fanatic of the video game medium, and an observer of the excitement around the increasingly powerful and creative video games of today. Whether we have the data or not, I will continue to explore the potential of the video game medium for creating vibrant, exciting, and engaging learning experiences that can be transferred to real-life situations. I appreciate the efforts of individuals like James Paul Gee and organizations like the Center for Game Science who have brought us to a point where the combination of video games and learning seems so promising. Because of their work, I believe we may be able to figure out how to consistently use the video game medium for exceptional transferable learning experiences, as long as we take our time and approach the matter insightfully.

Now is our chance as gamers, game enthusiasts, researchers, educators, and scholars to tackle the social challenges that impede advanced learning games from thriving. If we can do this, there is no telling what we will be able to do for learning through the video game medium. If, in the end, we fail to prove that advanced learning games are definitively helpful for transferable learning, *at least* we gave it our best shot. Even with the possibility of failure, I believe that the best days are indeed ahead for learning games that take significant advantage of the video game medium — as long as we do not give up on them.

REFERENCES

Atari, Inc. (n.d.). *Atari history 1972-1984.* Retrieved from
     https://www.atari.com/history/1972-1984-0

Connected Learning Alliance. (2011, April 20). *Why video games are such
     a force in learning, civics, and social innovation* [Video file].
     Retrieved from https://vimeo.com/22671352

DeLoura, M. (2014, Oct. 6). The White House education game jam. *The
     White House.* Retrieved from
     https://www.whitehouse.gov/blog/2014/10/06/white-house-
     education-game-jam

DiCerbo, K. (2015, July 19). Taking games seriously in education.
     Educause Review. Retrieved from
     http://er.educause.edu/articles/2015/7/taking-serious-games-
     seriously-in-education

Digital Media and Learning Research Hub [DMLResearchHub].
     (2011, August 4). *Games and education scholar James Paul
     Gee on video games, learning, and literacy* [Video file]. Retrieved
     from http://www.youtube.com/watch?v=LNfPdaKYOPI

Entertainment Software Association. (2015, April). *2015 essential
     facts about the computer and video game industry.*
     Retrieved from http://www.theesa.com/wp-
     content/uploads/2015/04/ESA-Essential-Facts-2015.pdf

Entertainment Software Association. (2016, April). *2016 essential
     facts about the computer and video game industry.*
     Retrieved from http://essentialfacts.theesa.com/Essential-
     Facts-2016.pdf

Gee, J. (2003). *What video games have to teach us about learning and
     literacy.* New York, NY: Palgrave MacMillan.

Graafland, M., Dankbaar, M., Mert, A., Lagro, J., De Wit-Zuurendonk,
     L., Schuit, S., … Schijven, M. (2014, Nov. 11). How to
     systematically assess serious games applied to health care. *JMIR
     Serious Games, 2*(2). Doi:10.2196/games.3825.

Higher Education Video Game Alliance. (2016, July 21). *HEVGA president Constance Steinkuehler to speak at event during democratic national convention on 'the future of video games, interactive media & play'* [Press release]. Retrieved from https://hevga.org/article_writeups/article-writeup-01/

Houghton Mifflin Harcourt. (2015, Nov. 19). *Carmen Sandiego returns: Houghton Mifflin Harcourt launches character's first-ever iOS app* [Press release]. Retrieved from http://www.hmhco.com/media-center/press-releases/2015/november/carmen-sandiego-returns

Lenhart, A. (2015). Teens, social media & technology overview 2015. *Pew Research Center*. Retrieved from http://www.pewinternet.org/2015/04/09/teens-social-media-technology-2015

Mathews, J. (2015, May 3). How video games could make our kids smarter and learning more engaging. *The Washington Post*. Retrieved from https://www.washingtonpost.com/local/education/how video-games-could-make-our-kids-smarter-and-learning-more engaging/2015/05/03/7690add4-ef87-11e4-a55f 38924fca94f9_story.html

Metz, E., Shilling, R., & DeLoura, M. (2014, Sept). Ed games week highlights the emergence of video games in education. *Homeroom: The Official Blog of the U.S. Department of Education.* Retrieved from http://blog.ed.gov/2014/09/ed-games-week-highlights-the-emergence-of-video-games-in-education/

Pappano, L. (2012, Nov. 2). The year of the MOOC. *The New York Times*. Retrieved from http://www.nytimes.com/2012/11/04/education/edli /massive-open-online-courses-are-multiplying-at-a rapid-pace.html

Perna, L., Ruby, A., Boruch, R., Wang, N., Scull, J., Evans, C., & Ahmad, S. (2013). The life cycle of a million MOOC users. *MOOC Research Initiative Conference.* Retrieved from http://www.gse.upenn.edu/pdf/ahead/perna_ruby_ bruch_moocs_dec2013.pdf

Serious Play Conference. (2015, June 29). Games for learning titles earn awards in international serious play competition [Press release]. Retrieved from http://seriousplayconf.com/june-29-2015-awards/

Sharples, M., Adams, A., Alozie, N., Ferguson, R., FitzGerald, E., Gaved, M., … Yarnall, L. (2015). *Innovating pedagogy: Open university innovation report 4.* Retrieved from http://proxima.iet.open.ac.uk/public/innovating_ pedagogy_2015.pdf

Sharples, M., McAndrew, P., Weller, M., Ferguson, R., FitzGerald, E., Hirst, T., Gaved, M. (2013). *Innovating pedagogy: Open university innovation report 2.* Retrieved from http://www.open.ac.uk/iet/main/sites/www.open.ac.uk.iet. main/files/files/ecms/web-content/Innovating_Pedagogy_ report_2013.pdf

Squire, K. (2011). *Video games and learning: Teaching and participatory culture in the digital age.* New York, NY: Teachers College Press.

SuperData Research, Inc. (2016). *eSports market report.* Retrieved from https://www.superdataresearch.com/ market-data/esports-market-brief

The Internet Archive. (n.d.). *Math Blaster (4am crack).* Retrieved from https://archive.org/details/MathBlaster4amCrack

The Strong National Museum of Play. (2016a). *The Oregon Trail: World video game hall of fame. Retrieved* from http://www.worldvideogamehalloffame.org/games/oregon-trail

The Strong National Museum of Play (2016b, May 5). *World video game hall of fame inductees announced* [Press release]. Retrieved from http://www.museumofplay.org/press/releases/2016/05/2688-2016-world-video-game-hall-fame-inductees-announced

The White House. (2011, March 8). *President Obama on education at TechBoston.* Retrieved from https://www.whitehouse.gov/photos-and-video/video/2011/03/09/president-obama-education-techboston

Toppo, G. (2015). *The game believes in you: How digital play can make our kids smarter.* New York, NY: Palgrave MacMillan.

U.S. Department of Education, Office of Educational Technology. (2015, April). *Ed tech developer's guide. A primer for software developers, startups, and entrepreneurs.* Retrieved from http://tech.ed.gov/files/2015/04/Developer-Toolkit.pdf

U.S. Department of Education, Office of Educational Technology. (2016, January). *Future ready learning: Reimagining the role of technology in education.* Retrieved from http://tech.ed.gov/files/2015/12/NETP16.pdf

Wink, C. (2013, April 3). How a Drexel professor will play Pong on the Cira Centre. *Technical.ly Philly.* Retrieved from http://technical.ly/philly/2013/04/03/how-frank-lee-got-pong-on-the-cira-centre/

Wright, M. (2013, April 9). MOOCS: A design question to consider. *The Huffington Post.* Retrieved from http://www.huffingtonpost.com/marcus-t-wright/moocs-a-design-question-to-consider_b_3033999.html

Wright, M. (2014a, July 9). "Why you should go play after reading this—a Q&A with Kevin Carroll." *The Huffington Post.* Retrieved from http://www.huffingtonpost.com/marcus-t-wright/post_7994_b_5568372.html

Wright, M. (2014b, July 16). Speak 180: Why is game-based learning good for today's youth? *The Huffington Post.* Retrieved from http://www.huffingtonpost.com/marcus-t-wright/speak-180-why-is-gamebase_b_5587850.html

MODERN LEARNING GAMES MENTIONED IN THE BOOK

Decisions that Matter. Carnegie Mellon University.
andrew.cmu.edu/course/53-610.

Fair Play. GLS Studios at University of Wisconsin-Madison.

gameslearningsociety.org/fairplay_microsite/

Spent. Urban Ministries of Durham. playspent.org.

Treefrog Treasure. Center for Game Science at University of Washington.
play.centerforgamescience.org/treefrog.

ABOUT THE AUTHOR

Marcus T. Wright is a higher education professional dedicated to the academic and professional success of college students. Using an approach that consistently connects the academic world with "the real world," Marcus advises and teaches students at all levels of college life. Currently, he is the Undergraduate Program and Communications Manager in the Department of Sociology at the University of Pennsylvania, a Pre-Major Advisor in Penn's College of Arts & Sciences, an Adjunct Professor in Communications at Holy Family University, and a Faculty Supervisor and Program Advisory Board participant for Cooperative Education in their Communications program.

Marcus is also a commentator on learning technologies, focusing on the intersection between learning, technology, and society in the digital age. He is particularly interested in the future and potential of these technologies for learning on college campuses outside of the classroom. His research and analysis on this topic has covered traditional online learning, Massive Open Online Courses (MOOCs), and game-based learning.

Marcus is an avid writer. He has contributed written works to *The Huffington Post*, *Entertainment Weekly's EW Community*, and several other online publications. Some of this writing spans the depths of pop culture. He has conducted written interviews with notable figures including former Super Bowl-winning NFL head coach Tony Dungy, violinist Lindsey Stirling, *Entertainment Weekly* correspondent Nina Terrero, social change agent and author Kevin Carroll, and musical ensemble The Piano Guys.

Born in Philadelphia, PA, Marcus graduated grade school from the historic Girard College (a 1st-12th grade private boarding school in Philadelphia for children from financially limited single-parent or guardian-led homes). He earned his undergraduate degree in Sociology from Rutgers University (New Brunswick, NJ), before returning to Philadelphia to start a high school sports multimedia coverage company, Varsity 365. After covering many of "the biggest games and best rivalries" in Philadelphia area high school sports, Marcus shut down Varsity 365 to return to the books. He then earned his M.S.Ed. in Learning Sciences &

Technologies at the University of Pennsylvania's Graduate School of Education (Penn GSE), followed by a Certificate in Penn GSE's Virtual Online Teaching program.

For more on Marcus and his work,
visit his website:

www.marcustwright.com

www.ingramcontent.com/pod-product-compliance
Lightning Source LLC
Chambersburg PA
CBHW060622200326
41521CB00007B/860